The Science Notebook

Teacher Education Edition

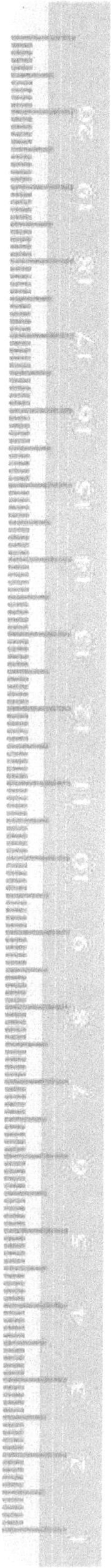

Science & Engineering Practices

Next Generation Science Standards

1. Asking questions for science and defining problems for engineering
2. Developing and using models
3. Planning and carrying out investigations
4. Analyzing and interpreting data
5. Using mathematics and computational thinking
6. Constructing explanations for science and designing solutions for engineering
7. Engaging in argument from evidence
8. Obtaining, evaluating, and communicating information

--NGSS Lead States (2013)

Common Metric Measurements

Listed in their category from small to large

Length or Distance: mm, cm, meter, km

The numbers on this ruler are in cm.

The spaces between are mm.

Mass: gram, kg

Volume: ml (cm^3), liter, m^3

Energy or Work: joule, kJ

Force: newton

ISBN: 978-0-9988127-2-4

Independent Variable, LLC

Table of Contents

Date	Topic/Activity	Pages

Date	Topic/Activity	Pages

Special Features and Readings

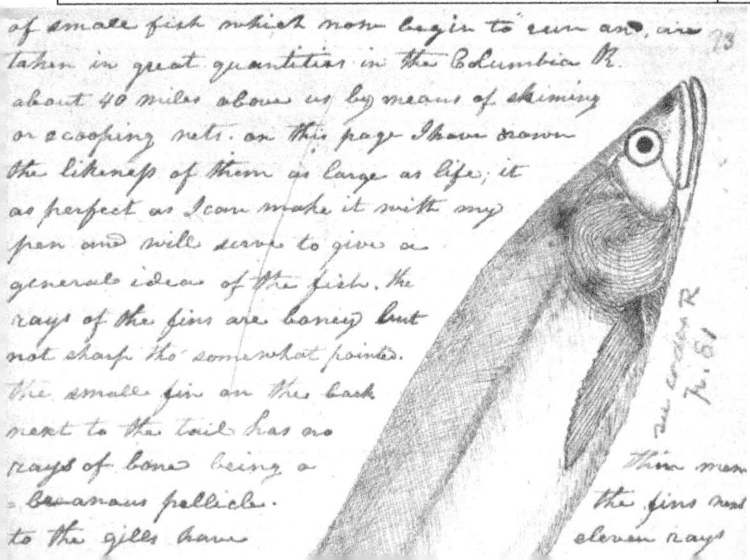

As seen on the left, this journal entry, recorded by Meriwether Lewis in 1806, describes the eulachon fish in Oregon. Lewis was a better technical drawer than his exploration partner William Clark who was a better cartographer.

Science Notebooking

With good reason, science notebooks have become common in school science classes. They provide structure for notetaking and recording observations. They provide opportunities for reflection and to deepen thought, through acts of writing and concept mapping. Notebooks also model the actions of scientists and engineers as they work.

This notebook has been created for you, the preservice teacher. It will help you deepen your understanding of practical aspects of notebooking and help you think about the great potential they have for you as a teacher and a learner.

Guide for this Notebook

There are three sections for this notebook, each numbered independently. The front section has the table of contents, science and engineering practices, and common metric units. This current section is the main body, with normal Arabic page numbers. It contains dotted pages and blank pages for you to write, draw, table, or graph as part of the learning process. Add the date and topic to each page and record these in the Table of Contents. Readings and embedded activities promote learning and will make your notebook a cherished possession. Page VI of Section 1 has a table of these features. After you finish an activity, enter the date of completion. The symbol ⊡ is used to help you see certain action items. Consider adding a check mark to the boxes as you complete them. You are encouraged to reflect as learners and future teachers. After class experiences, you will draw a fanciful line-of-learning (LOL), and then write your thoughts.

The main body ends with the Concept Index, where you keep track of scientific and educational concepts. The third section has the Concept Clavis, which is where you list, describe, and draw the concepts. When you have a concept, put it in the correct alphabetical box in the Index, with the main page number and Clavis page number after it. Then you list, describe, and draw it in the Clavis.

Starting Activities

On the inside cover, ⊡ it is suggested that you write your name and provide a note and contact information in case you lose this notebook. ⊡Semesters can get long, so you are asked to provide motivation for your future self. Turn to page 22 and write a note to your future self; something to motivate or inspire you. Finally, read the Personify Science section (p. 2), and think about this. Later come back to this and write you narrative. ⊡ Then practice using the Index (p. 36) and Clavis. You will find "personify" has been written for you in the Index box with the P-words. Afterwards, move forward one page from the index to the Clavis and ⊡ write "personify" as the first word, then describe and draw it.

Personifying Science

Date:

Pretend Science is an actual person—maybe a friend, maybe a foe, maybe someone in between. Write either a (a) narrative or (b) letter to Science, no longer than two pages, about who Science is and your relationship with he/she/it/they. Some questions to guide you might include:

How long have you known each other? What does he/she/it/they look like? How does he/she/it/they act? How has your relationship with Science changed over time?

Activity adapted with permission from Grimes, Reid, & Smith-Walters, 2020.

Chemical Equations

Chemical equations show the **reactants** (left side of arrow) forming **products** on the right side. In the equation below, methane combines with oxygen to yield carbon dioxide and water.

$$CH_4 \quad + \quad 2\,O_2 \longrightarrow CO_2 \quad + \quad 2\,H_2O$$

The atoms in the equation can be represented by spheres and are shown below.

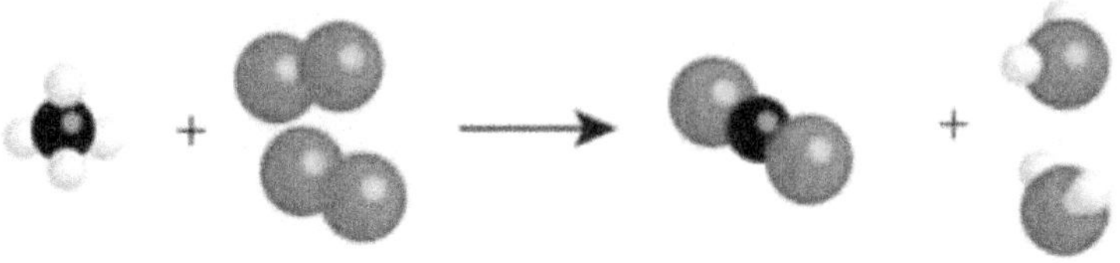

Life Equations

☐ Now, make four fun or deep Life Equations using the chemical equation format.

Here is an example:

Desire + Urgency ⟶ Action

☐ Products and reactants have been added to your Concept Index.

Please add them to your Clavis, where you will describe and draw them.

What is it?

Why do you think that?

Why are some drawn bigger?

The above is a representation. The real thing is considered **realia**. Think about a memorable time seeing the real thing. Describe it.

What is a ray? These straight beams of light travel from luminous bodies.

☐ What are you grateful for? Reflect and write between the rays.

What we observe is not nature itself, but nature exposed to our method of questioning. ~Werner Heisenberg, theoretical physicist

What is it?

Why do you think that?

Graphene: A wonder material with a 2-d, single layer of carbon atoms.

It is 200x stronger than steel, lighter than paper, flexible, and conducts electricity

Chlorophyll Painting

Using fresh green leaves from nature or the grocery store. Roll and smash them between your fingers to release chlorophyll. Then use them as paint and brushes to add foliage to the tree above. If you are inspired, smudge soil to color the ground below the tree and use natural pigments to add fruit or flowers to the tree. Look up the role of chlorophyll and it to your Concept Clavis.

Add an inspiring note to your future self at the start of the semester.

Failure is an option here. If things are not failing, you are not innovating enough.

–Elon Musk, Entrepreneur

Stain Station

Write Stain Name

Fill the above boxes with at least 14 different stains, writing the substance name below each. Consider adding a blotter page behind this page, so the stain doesn't diffuse into other pages.

We see color, for example red, because the stain absorbs all other colors but reflects red light rays back to our eyes. In chromatography, scientists separate pigments by having them dissolve and rise up (through capillary action) paper or other materials. Stains and pigments are usually composed of many colors, which separate in chromatography.

5-E Model for Inquiry

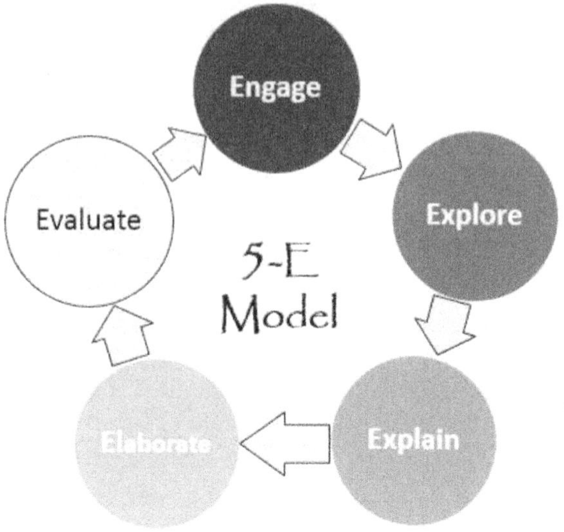

The 5-E Model (Bybee, et al., 2006; Rillero, 1999), shown in the diagram, enhances inquiry approaches. It facilitates learning by encouraging students to actively build their knowledge.

Engage: This phase creates interest and generates curiosity in the topic of study. It tends to be the shortest phase. It could be as simple as: Hold up a working pulley. Ask students where they have seen them used. Then inform them "today we are going to explore how pulleys work."

Explore: During this phase, students are provided opportunities to work together without a lecture by the teacher. In science, this is what is typically thought of as a hands-on activity, laboratory experience, or experiment. Teachers provide initial directions and then serve as facilitators with prompts and questions to help students observe, question, think, and process. Science exploration presents opportunities to record observations, data, and ideas. Activity sheets can be useful to clarify directions and record results.

Describe a Science Hands-On Activity you did in School. Try to think of one early in your school years.

⊡

Explain: Teachers initially encourage students to explain concepts in their own words and listen to other explanations. Questions and prompts help students think and share. Encourage students to use observations and data in their explanations. Later in this phase, the teacher introduces vocabulary, descriptions, and explanations, building upon the shared experiences. Towards the end of this section, teachers clarify and review explanations.

Elaborate: During Elaborate, students apply concepts and skills in new situations and use concepts and vocabulary developed. This can be a similar hands-on activity as was done in Explore. It could also be a different activity, such as making a concept map or foldable. Key elements of Elaborate are (a) it is a student-centered activity (not a lecture) which, (b) deepens understanding and/or abilities.

Evaluate: This phase provides additional experiences for students to demonstrate learning and abilities to apply concepts or skills. Teachers give feedback, which provides another learning opportunity for students. ⊡ In the diagram above, <u>choose one color and color the Evaluate phase and the arrows between all phases with it</u>. This is a reminder that evaluation occurs throughout the process.

Alternative Conceptions

Graduation bells ring as Harvard's seniors, wearing caps and gowns, are asked the reason for seasons. The most-popular view is well articulated: Seasons happen because the Earth travels closer to the Sun in summer than in winter. 21 of 23 interviewees didn't have accurate conceptions about causes of either the seasons or phases of the moon. **Alternative conceptions** are ideas that differ from scientifically accepted ideas.

Sometimes called misconceptions; alternative conceptions (ACs) is preferred because ideas are not always 100% right or wrong; there are shades of being right as students progress towards scientific conceptions. We want to foster these journeys. Second, views that differ from scientific views may not be wrong. We know scientific ideas change over time (maybe you view Pluto as a planet). Perhaps more evidence will make the idea of a scientist more broadly accepted by the scientific community. Perhaps the ideas are from a spiritual or religious perspective, where what counts as evidence for truth differs. AC is not just PC; its use reflects that people may have a different view but it is not necessarily wrong.

ACs are resistant to change. Well-schooled adults explain that we see phases of the moon because the Earth blocks light from the Sun. They explain that the mass of the heavy mesquite log came from the soil. As children age, logic is shown with ACs. A parent fertilizing a plant tells a child he is feeding the plant. The label of the fertilizer says "Plant Food." So, it is logical to think the tree obtained its mass from the soil. The scientific conception of obtaining most of its mass from carbon dioxide in the air, through the process of photosynthesis is far less obvious.

People don't like to be wrong. Therefore, lectures or readings may not change ACs. A key step is making students question their thinking. For instance, you could ask the Harvard graduates, "If being close to the Sun causes summer, how can you explain why it would be summer in Massachusetts but winter in Buenos Aires?"

Giving students many opportunities and perspectives on a topic can help them develop scientific understandings. **Discrepant events**—experiences that have a surprising outcome—can be effective demonstrations or hands-on experiences to promote the accommodation of scientific ideas.

⊡ For the following list of ACs, write the scientific conception below it.

Energy is created during processes such as combustion.

Light reflects off smooth surfaces like mirrors but not other surfaces like paper.

A thermometer wrapped in a blanket will warm up.

Dissolving and melting are the same.

Concept Maps

When we think of something, such as a favorite animal, we think about attributes and qualities based upon our experiences, such as being furry, warm nosed, and caring for its offspring. We rarely think with formal, standardized definitions. Providing experiences that help students develop understandings of science concepts is an important goal. Concept maps (CMs) can help achieve this. They differ from other graphic oganizers in the following two ways: (a) hierarchical organization and (b) linking words that join all concepts.

Concepts are put into a shape, such as an oval or rectangle. Arrows, with embedded linking words, join the concepts. The concepts are arranged as a hierarchy, with the most general words at the top, becoming more specific moving down the map. Linking words connect concepts together. When you read one concept and the linking word and then the next concept, it forms a proposition, which is like a sentence. The short CM below shows both the hierarchy and linking words.

Concept mapping is a powerful learning tool because it makes creators think hard about the concepts and their relationships (Novak, 1990). Organizing a hierarchy takes thought; write down the concepts you want to use and then consider which are more general and should go at the top of the map. As you start making the map, think about connecting the words to form propositions. Sometimes the linking words just fit, sometimes you will need to adjust the wording of the concepts. It is best to not repeat the same concept on a map but to have links from various places to the concept. Crosslinks connect concepts in different areas of the map; they display deep thought about the topic.

CMs can effectively promote and assess learning. A good starting place is to have students add missing concepts or linking words to maps. ⊡ For the CM below, think about logical additions and add a concept (A), a linking word (B), and then add a new concept, arrow **and** linking word in the region of (C). For example, you might have an example of an element (a concept) coming off Elements, with a proper linking word.

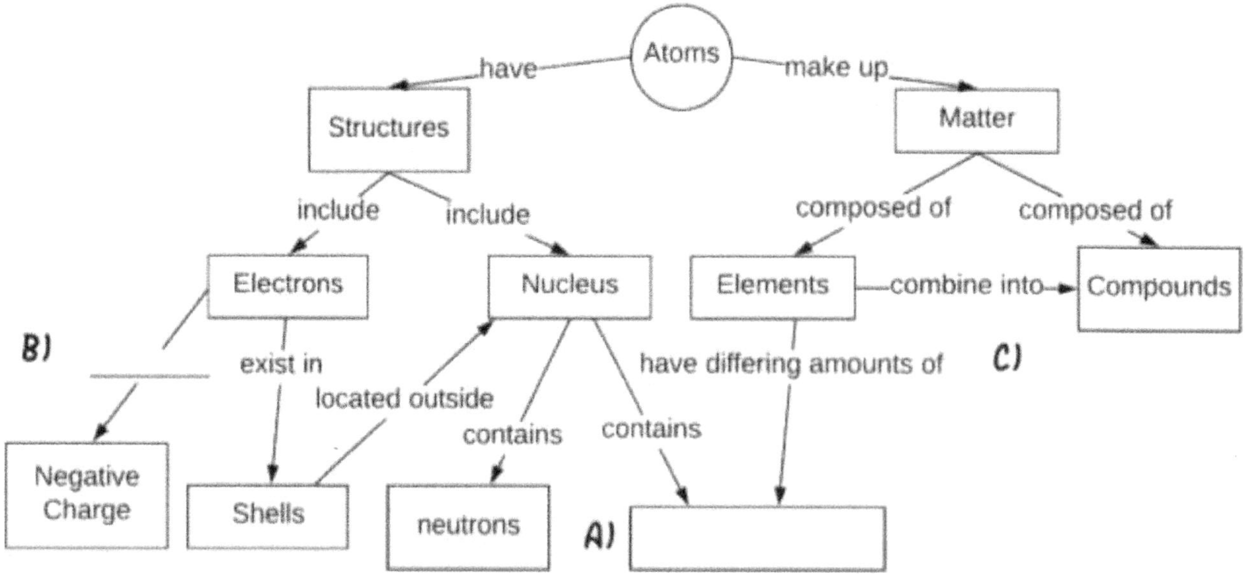

A Vision of Myself as a Future Teacher of Science

In the space above, draw yourself as a future teacher of science. In the area below, (A) write what you are doing. (B) Explain what your students are doing.

Concept Index

Concept Clavis

A clavis is a key or glossary that assists in interpretation. Enter concepts chronologically; words at the beginning of the semester start the list. Describe what the concepts mean to you. Draw a picture of it.

Concept:	Describe:
Draw:	

Concept:	Describe:
Draw:	

Concept:	Describe:
Draw:	

Concept:	Describe:
Draw:	

Concept:	Describe:
Draw:	

Concept:	Describe:
Draw:	

Concept:	Describe:
Draw:	

Concept:	Describe:
Draw:	

Concept:	Describe:
Draw:	

Concept:	Describe:
Draw:	

Concept:	Describe:
Draw:	

Concept:	Describe:
Draw:	

Concept:	Describe:
Draw:	

Concept:	Describe:
Draw:	

Concept:	Describe:
Draw:	

Concept:	Describe:
Draw:	

Concept:	Describe:
Draw:	

Concept:	Describe:
Draw:	

Concept:	Describe:
Draw:	

Concept:	Describe:
Draw:	

Concept:	Describe:
Draw:	

Concept:	Describe:
Draw:	

Concept:	Describe:
Draw:	

Concept:	Describe:
Draw:	

Concept:	Describe:
Draw:	

Concept:	Describe:
Draw:	

Concept:	Describe:
Draw:	

Concept:	Describe:
Draw:	

Concept:	Describe:
Draw:	

Concept:	Describe:
Draw:	

References

Bybee, R. W., Taylor, J. A., Gardner, A, van Scotter, P., Powell, J., Westbrook, A., & Landes, N. (2006). *The BSCS 5E instructional model: Origins and effectiveness.* Colorado Springs, CO: BCBS.

Grimes, Z.T., Reid, J.W., & Smith-Walters, C. (2020). Personification as a Reflective Tool to Examine Pre-Service Teachers Attitudes Towards and Conceptions of Biology. Presentation at the International Conference of the Association of Science Teacher Education.

NGSS Lead States. (2013). *Next Generation Science Standards: For states, by states.* Washington, DC: The National Academies Press.

Novak, J. (1990). Concept mapping: A useful tool for science education. *J. Res. Sci. Teaching, 27*(10), 937-949.

Rillero, P. (1999, June). *Raphanus sativus*, Germination, and Inquiry: A Learning Cycle Approach for Novice Experimenters. *Electronic Journal of Science Education, 3*(4).

Acknowledgements

I thank my colleagues in science education at Arizona State University, especially Mo Walters, Amy Bartlett, and Dianne McKee for their contagious enthusiasm for science notebooks. I thank my wife (Kim) for helping make this more readable and enjoyable, my son (Hy) for the cover design, my niece-in-law (Laura A. Rillero) for the careful review, and master science teacher Brooke Mullins for pre-publishing feedback.

Waste Page

Isaac Newton kept notes in his repurposed "Waste Book" that would later be developed into calculus. Make the back cover into your two-dimensional waste page. Save items from newspapers or magazines, fruit or veggie stickers, restaurant receipts, candy wrappers, wrapping paper from gifts, bills you paid, and anything that you can flatten that will provide memories. Glue them inside the back cover to make a collage.

Made in the USA
Monee, IL
08 November 2026

67378664R00031